"Steve is an excellent business strategist with a unique ability to bring the right level of technology to execute business initiatives. His aptitude for understanding what's broken in the business and how to apply technology to fix it is unparalleled. Steve's rather unique upbringing, education, and experience enable him to help companies identify value opportunities and achieve business objectives faster than anticipated. Steve and I worked in the robotic process automation space where I witnessed first-hand his genius in teaching and applying RPA to help clients understand the technology, document processes, and implement RPA to drive successful business outcomes. When I launch my next start-up business venture, I want Steve to lead the build out of my technology infrastructure and ensure alignment with our business objectives."

- Bill Varon
Account Executive

"I have known Steve for nearly 20 years now. We met while both working at what is now AT&T. Steve worked in Accounting & Finance and I worked in Operations. We collaborated on several cross-organizational projects and he excelled at forecasting and planning activities. Using technology and deep mathematical skills, Steve would forecast revenue and usage numbers. His forecasts were so accurate that he was able to come within 1% of actuals, weeks if not months ahead of reporting cycles. Steve was also part of the NCR Teradata Team, where he helped test and design the data collection and reporting systems. His reporting system design was light years ahead of our existing reporting. It provided highly valuable insights into the organization that never been achieved before, and he was the basis for driving broad organizational and process changes. He quickly became the go-to guy for reporting and financial data. Steve is one of the most capable and intelligent professionals I have ever had the pleasure of working with, and 20 years later he is still my go-to guy for technology and automation."

- Jason Fleming
Allstate – Agency Owner & Serial Entrepreneur

AUTHOR'S ACKNOWLEDGEMENT

I want to thank my wife Tammy for supporting me in getting this book across the finish line and being patient with me for all the hours spent focused on this effort in addition to supporting our family. And to my oldest brother Ernie, who always encouraged me to stay in school and do my best.

I also want to thank my dear friend Ann Hurley, who was my sounding board around the digital transformation process and project management integrations.

STEVE PALOMINO

PREPARE
FOR
CHANGE

Designing a Digital Transformation Strategy

By Steve Palomino, CPA, CIA, CITP, CGMA

TABLE OF CONTENTS

Introduction

When I became a CPA, I was required to take continuing education courses each year. The courses could be long and boring and after I was done, I would get my credits, but lose several very precious hours of my life and not learn much. As a practicing accountant, I became very involved with databases to support my work, and I wanted to learn how to execute advanced commands in SQL, Visual Basic and several other languages. I took many courses seeking knowledge, but often found myself teaching the course, because I had more experience at coding than the instructor. There was a great deal of theory or basic knowledge in the training, but it was obvious to me that most instructors did not have real-world experience. That is why my goal with this book is to provide the reader with actual hands-on tools to successfully implement a digital transformation strategy that will help your business be more profitable and nimbler. I developed my tools based on my experiences as both a corporate accountant at a Fortune Fifty communications company and as a consultant for the last 15 plus years. If you are a business owner, operations/business manager or C-level executive, this book is for you. After reading this book, it is my sincerest hope that you feel the time you spent is worthwhile, that you better understand digital transformation strategies, and that you can take a practical approach to implementing various strategies to drive continuous growth and profitability.

CHAPTER 1

The Caveman and the Spreadsheet

Thousands of years ago, humans struggled to survive in very harsh living conditions. Early humans were fragile in comparison to other animals that roamed the land at the same time. Staying alive was a major challenge for our early ancestors. But somewhere along our development, we, as a species, developed tool-making capabilities. No one knows if it was by accident or early genius, but somewhere along the line, a caveman picked up a pointed stick, tossed it at what they hoped to be their next meal, and lo and behold it worked.

As they say, the rest is history. Like early man, early business owners - and eventually accountants - did their best to manage an unmanageable business environment. Early Chinese invented the abacus to help them count lots of "things," and perhaps this is why the silk trade was such a successful operation. Later, the Italians invented the double entry system and in the 19th century, accounting as a modern profession took off.

From the time the double entry system was developed by Benedetto Cotrugli in the mid-1400s, until the 1990s, not much changed in the way business was managed. Large amounts of paper were used to track business affairs. Individuals tasked with "counting" things used paper, pencils (later adding machines) to track all business activities. These business counters later become our modern-day accountants. Their job was to observe items, write them down, have someone check their work, and then log the amounts in a final "journal" of business activities.

Then along came Bill Gates and IBM and the world of doing business was forever changed. It was the invention of the spreadsheet. This electronic innovation, like the cave man's pointed stick, was an inadvertent development. Rumor has it that at first, neither Bill Gates nor IBM really knew what to do with the spreadsheet[1]. I remember reading that the CEO of IBM thought it would be a good tool to keep track of his wife's recipes. So the spreadsheet became the modern business person's "pointed stick."

Well, this "new stick" turned age-old processes from a manual, physical set of actions and tasks into a virtual exercise. The rest is accounting and business operational history. From 1996 forward, the spreadsheet has

---

[1] https://www.pbs.org/nerds/

been the cornerstone of business management. This tool has been used to fill in every technology gap that exits, and long outlived its time.

But what really changed in the way business was managed? Were business and accounting processes executing any differently in a virtual environment, versus a paper-based environment? I assert that very little changed. Processes were and remain mostly a manual effort. Now they take place in a virtual environment, but they are not much more efficient.

While physical manufacturing has been and continues to be transformed by process improvement, "Lean Six Sigma" and robotics, virtual labor activities have experienced very little change.

This book will explore how to identify your true business needs regarding people, process and technology, how to determine the best course of action to meet your business needs, how to successfully implement a process change, and how to implement a technology to make your business more efficient and profitable. Finally, you will receive practical, real life instructions on how to successfully implement changes based on a digital transformation strategy, and I will provide guidance on how to build and manage that strategy.

CHAPTER 2

Understanding your Business Needs

Over the last 20 years, I taught numerous courses for the American Institute of Certified Public Accountants, conducted numerous webinars, and responded to interviews about how to best approach and take action in the ever-changing business environment. My contributions include *An Overview of Data Management, Business Intelligence Value Proposition Tool Kit Case Study, Big Builders*. Additionally, I was a contributing author for the AICPA's *Approach to Business Solution Implementations*, co-author of *A Practice Aid for Records Retention* with case study citation, *Improving Decision Making in Organizations, Unlocking Business Intelligence*. Most recently, I co-authored a multi-part series, titled "Robotic Process Automation Fundamentals for Accounting and Finance Professionals Certificate" for the AICPA.

As I am writing this book, a unique situation has arisen in our business environment. Well, unique to everyone born in the United States after 1920s. In March of 2020, a world-wide pandemic made its debut. This pandemic made doing business particularly challenging because companies were required to immediately respond to many different types of external forces. Each force had to be considered, whether driven by the government, supply chain issues or changing customer needs or habits, and past experiences were used to understand current challenges.

But what if your business was not prepared with complete, accurate and timely historical information? What if your company lacked the internal resources to manage and analyze new data trends against old data trends? Or worse, what if your business lacked the ability to adjust to the *changing* business trends?

All the issues mentioned above happen all the time, but are amplified in times of rapid change or economic shock. What we're seeing in today's business environment underscores how rapidly and drastically the business environment can change.

Preparation is the key to managing your response in a quickly-changing business environment. In today's scenario, the questions are, "Does your company have the resources to support 90% remote working? Does your business have a culture of collaboration? Does your business rely on a third-party IT group that is able to scale and meet your business needs? Does your business still rely on paper and the need for manual signatures?" All these issues - and many more - need to be considered when determining common business needs.

What are the common business needs?

1. The need to work from "one version of the business truth"
2. The need to consolidate data into one source of information
3. The need to get better information, faster
4. The need to reduce back office costs
5. The need to add or maintain headcount
6. The need to control financial processes and reporting
7. The need to control operational processes and reporting
8. The need to do all the above while focusing on your core business
9. Most recently, the need to be able to work remotely

When we look at the business needs noted above, we can start to see how to group them into buckets. Once you understand the buckets, we can then discuss what solutions might be applied in and around each bucket.

Let's take a more detailed look at each one of these needs.

Bucket One

Data – obtaining data, transforming data, matching data, consistent data

- The need to work from "one version of the business truth"
- The need to consolidate data into one source of information
- The need to get better information, faster

One version of the truth

This is the same for small, medium and large business entities. No business structure appears immune from the need to know "the right number" When I worked for a very large telecom company based in Texas. I sat through hundreds of meetings in which Presidents, Vice Presidents and other senior leadership argued about the numbers that they were being measured against. Each business unit group would bring in their own accountant to describe their numbers and argue that "their version of the truth" were more accurate than the ones they were being measured against. This dance went on for years and it never seemed to end. As the company grew via acquisition, more and more systems were added to the mix, causing even more confusion and dissent.

This ever-moving target made it very difficult to make rational business decisions. For example, large amounts of money were spent on various ad campaigns to grow product market share or to introduce our services in a newly acquired market. When I reviewed the revenue generated by the ads, it might show little or no change for the market or for the business unit that had kicked off the marketing campaign. After my review, a different business unit finance team in the same organization would review data from their own "data source" and find revenue growth. This led to a data fire drill. My team was immediately tasked with going into all the systems to see just what was driving the shift in the data. Field visits took place and thousands of hours were spent trying to prove or disprove that the campaign was effective. The final outcome was often analysis-paralysis, and more time spent arguing the numbers than how, when and where to apply a business strategy.

But my experiences didn't end there. When I moved into a consulting position, I found that small and medium sized businesses faced the same issues. At a mid-sized transportation company, they used a legacy general ledger system. They also extensively used Excel. The result was several sets of "books" used to manage the business. According to some data, the business was trending toward growth, and according to other data it appeared the business was shrinking. This was all caused by the lack of one uniformed set of data.

I later had the chance to work with a very small company that used a small online GL application to manage their books. Because the business lacked any real controls over journal entries and did not follow Generally Accepted Accounting Principles, the use of credits and discounts were shown as rebates. This approach made it very difficult to price products, because no trending could be seen in the data. It was hard to see what was driven by business activity and what was driven by internal accounting activities.

When data is not consistent, consolidation can make things even more challenging.

Consolidating - One Source of the Truth

When viewing data separately, it is much easier to manage and identify discrepancies in the data. Systems can be reviewed, reports and underlying data tested. However, when data is rolled into a single reporting environment, it becomes very difficult to determine what is driving the differences.

Take a standard GL system that is used in mid-market type businesses. The database may have feeds from such external applications as Concur

Expense, payroll data from external sources such as ADP, and sub-ledger data from various sources. I have observed many, many cases where companies are growing and adding new businesses, or acquiring new businesses, and the data is rolled into a main reporting or consolidation system. Accountants review the data, usually at the end of the month, and what they find are increases in account balances that don't make sense or decreases that don't appear to follow normal business trends.

One such business I worked with sold various types of office equipment. The business grew via acquisition of smaller businesses and relied on the companies' reporting systems for a period of time, then rolled the data into a larger system. This is not an uncommon business practice, but without proper oversight and management, data can be incomplete, or from the wrong time periods. Data may be acted upon by external forces, people or technology, and ultimately end up in the consolidated environment. When this occurs, consolidated balances become mottled with bad data and information cannot be relied upon without a great deal of review and manual oversight. For instance, at this office equipment sales company, it became necessary to review detail data line items and to continually move between summary data and detail data, so that customer payments could be applied to the correct customer account balance. Accomplishing this task required them to substantially increase their headcount. For this business, the challenge was very difficult when it came to hard equipment, but when subscription supplies, usage and leases are all added into the mix, one can see how complex and cumbersome the entire reporting process becomes. There is most certainly not just "one version" of the truth.

The Need to Get Better Information Faster

Data is just that, data. Zeroes and ones are strung together to make a meaningful character that reflects the outside world. Like right now you are reading a page that was produced electronically. To the PC I used to type this book, each letter is merely a series of bits and bytes. Once strung together, it becomes a letter and the letters become words or better, information. But we just don't want data, we want information and not just any information, but information necessary to run our business successfully. The information must also be available in time for business leaders to use in making decisions.

I had a professor many years ago who taught me about data bases and reporting. Technology was not as sophisticated as it is now, but it is amazing how we still run into the same issues. My professor told me.

"When it comes to speed of getting information, would you like all the answers after your test or a really good idea of what is going to be on the test, before you actually take the test?" What he was saying, and kept repeating, is that it takes too much time to get perfect information. And depending on the grade you seek, the more information you have early on, gives you the opportunity to prepare for an event.

The point above is this – you need to have enough information to be prepared to take the "business test," but not wait until the time to act has passed. Over all these years, data didn't get smaller and easier to use, but truly increased in volume and scope. Organizations become overwhelmed with the amount of data that is available.

Case in point. When I was at the large telecom company doing reporting all those years ago, we only had a few systems for reporting. We had our COBOL-driven billing systems. We had mini computers, called AS400's and then we had "Dunn & Bradstreet" as our general ledger systems. Our GL system put out very little data and it was mostly summarized. The data produced little useful information so I spent most of my time trying to analyze revenues, expenses and products sold by the business with very little useful, summarized information. But I did have more time to focus on researching the underlying data that appeared in the summarized reports. I was then able to turn that data into useful information. My time was spent on analysis and not just producing reporting packets and consolidation reports.

As the company grew, I ended up spending more and more time on reporting and less and less time on understanding and analyzing the business. The information I provided was less and less valuable at the reporting level, and it took more time and resources to produce reports. The business had much more detail to review. More detail on product information. More detail at the reporting center level...but we had less and less time to check and review the underlying data.

The growth of data became so problematic that an entire new consulting business line developed. That new consulting business was focused on shortening the closing and reporting cycle. Some businesses had gone from needing several days to close their books with limited resources, to needing large teams that took up to 3 weeks or more to close the books and report.

We ended up with less information, a longer wait to get that information, and more effort to obtain it.

Bucket Two

Timely and accurate information – information that is obtained in time to be useful in making quality business decisions.

- The need to control financial processes and reporting
- The need to control operational processes and reporting
- The need to do all the above while focusing on your core business

You've learned about better information that one gets faster. Now we move on to the other side of that coin which is accurate and timely information about business operations in the form of reports and other mediums.

The Need to Control Financial Processes and Reporting

There is a very, very old saying, "haste makes waste." In our attempt to turn data into information, to meet ever-changing business needs, we rush to produce as many reports as possible. These reports are generated by various parts of the business, and depending on the size, you may end up seeing reports produced by all parts of the business.

In one case, I was working in a corporate finance function where I was responsible for revenue assurance - making sure customers were being billed and revenue was not leaking out of the business. I would also see usage and traffic reports from operations, as usage drove much of our revenues. I could see that the number of clients was decreasing daily and that usage continued to drop. Each day, almost like clockwork, I could see that customers were leaving. I started to research the losses to make sure I was not seeing customers moving from one product to another product, or to another part of our business. The terms we used in the industry were churn and cannibalization.

As I began my research, I started to see reports being generated from various other teams in the company. Then I started to get calls from various VPs and other business leaders wanting to know why revenue was dropping. They were being told by a marketing team that calls from customers were up, but they were not seeing the revenue increase. So their thought was that a billing issue must be at play. In their eyes, there was a vacuum of information that could only be solved by producing reports.

The production of all of these reports led to a significant amount of work for all reporting teams. A great deal of confusion and consternation then followed, as leaders began to argue about their interpretations of the results. Once I and my team had a chance to execute the detailed research, we found that calls were indeed up across the business.

Unfortunately, the calls were to cancel service or to request reduction in monthly billing charges. We also found that our competitors were offering the same products and services we were providing at up to a 50% cost reduction.

The lack of controls in the reporting process not only resulted in valuable time being lost that could have been used to address the immediate need, but also damaged the reputation of reporting teams across the company. It was not that the reporting teams were malicious or incompetent. It was just that there were no controls over reporting - what should be contained in reports, who should disseminate reports and most importantly, what quality controls should be applied before reports are released to the business.

The Need to Control Operational Processes and Reporting

As the example above illustrates, it is imperative that controls exist over operational processes and reporting. Not just from a publicly held company prospective, but from a standard business management prospective. If there are no controls around operational processes, such as how customers are tracked, what metrics are being applied to customer tracking and who is responsible for such reporting? In this case, we also found that codes were not being applied to change orders that "removed" the customers for tracking applications. This minor but significant oversight resulted in a large number of customers sitting in cyber limbo. Next, we found that the operational teams had not assigned responsibility and accountability to tracking "ins and outs" for customers. The focus was on the actual provisioning of the customer, which was great. But who and how to keep track of customers was never addressed. This lack of ownership and accountability resulted in an information "free-for-all."

The Need to Do All of the Above While Focusing on Your Core Business

As business becomes more complex, it becomes a real challenge to manage your business and attend to what brings money into the business.

Lost in all of the reporting, reviewing and discussion are the customers' needs. Not just internal users of business information, but your customers. The downstream impact of the operational failure noted above was that when orders were finally cleaned up, it was discovered that a great deal of revenue had never been collected related to early termination fees, usage contracts and a myriad of other terms and conditions. When the business decided to bill for unbilled items and take other collections activities, a significant number of customers were adversely impacted. The effort was caused such a stir, that local news outlets began to carry the story.

Not being able to focus on the customer and the core business resulted in negative outcomes that took months to overcome. We exist to serve our customers – *without our core business, we don't have a business*.

Bucket Three

Resources, resources, resources – you don't get resources without the trailing expense, and the greatest expense related to most resources is salary. This is most evident by looking at personal incomes. I chose to look at one example from economagic.com. This is one of the best sites I have found when it comes to providing great, reliable sources of data.

I chose to pull the Mideast, which includes such places as New York, but also includes such places as Pennsylvania. Below is the graph[2]:

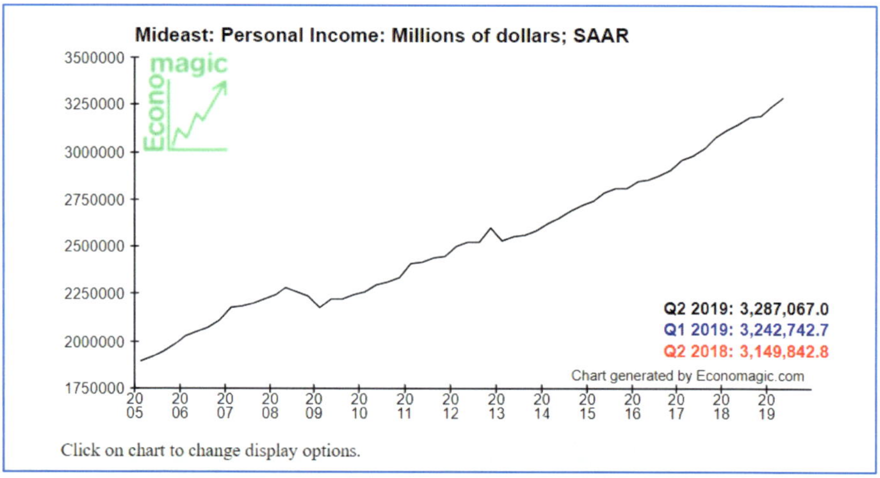

What the data indicates is that since 2005, on average wages have grown. This means that regardless of ups and downs, it will continue to be more expensive every year to add resources. In times of plenty, these costs will most certainly increase. In times of upheaval, we most certainly will see a drop, like the COVID-19 (Wuhan-originated virus) impacts in 2020. But after a few years, the capitalist US economic engine drives forward. This will ensure that quality, technical resources will be more and more expensive each year and harder to find, hire and retain. We also see that times of economic shock result in a decrease in the labor supply, because

[2] http://www.economagic.com/em-cgi/data.exe/beapi/fips92

many individuals who can retire often do so and never return to the workforce.

The Need to Reduce Back Office Costs

Back office expenses have long been a target of expense cutting measures. Some companies moved away from paper years ago to reduce costs. Others reduced the amount of office space used and eliminated such things as gym memberships, free coffee and such. But the big focus was on technology enablement. Applications sprung up all over the place, to help manage and lower expenses. Ten years ago I don't think anyone had ever heard of "Expensify" and applications like "Concur" were mainly used by very large companies with large numbers of employees. Offices managers also began to look at how infrastructure costs, such as high-speed internet, could be managed.

In 2005, I worked with a small lender who had only one office. For the small fee of $1,500 per month they received a bank of telephone lines via DS1 line service. DS1 is able to transmit up to 24 multiplexed voice and data calls over telephone lines. But this cost was really high and required expensive equipment that was leased or purchased from the phone company. So organizations began to look around and see how they could reduce this expense.

Along came the "internet of things" and the IP telephone became a standard staple in many offices. This resulted in significant cost savings. In some cases, IP phone service and related internet service can run as little as $100 per month. This shift in technology solution demonstrated for organizations that changing systems or leveraging new technologies could help to significantly reduce back office expenses. There are many, many more examples, but I think you get the idea.

The Need to Add or Maintain Headcount

Crashing against the goal of reducing back office expense is the need to add or maintain human resources. To address the need to lower costs and attempt to reduce headcount, companies began implementing many different reporting systems, and computerized supply chain tools. Gains were observed and some costs did drop. It was not enough, but it was a strong start.

Go back in time to around 2003. When I was at a large telecom company, my reporting team had approximately 15 accountants. We produced reports and struggled to meet the demands of the business with the resources we had. The business began to reduce headcount by not backfilling retirements and other voluntary resignations. Leaders decided to invest heavily in a data warehouse and other technologies. Reports were reduced and ultimately, just a few team members remained. Today, there are only a few individuals performing the tasks that 15 people used to do. However, this transition took years to accomplish, and at a hefty cost.

So that is what we observed. The cost of automation ultimately helped business save money, but it turned out that the payback period exceeded 5 years. As technology changed, the business was forced to continue investing in the legacy application or to upgrade applications to remain Microsoft operating system compliant. To complicate the cost savings, some applications were purchased by new software vendors and became obsolete, rendering the initial investment a total loss.

The Need to Work Remotely

For years, I was told by my employers that the idea of working remotely was just a bad idea. The leadership of most companies did not support a remote workforce. Then a few things changed. The telecom company I was working for started to see the importance and efficiency of working remote, and began using that to help drive sales. The idea was that businesses could save money on real estate and workers could be more productive. This was music to my ears. At least in the telecom industry, working remote was an acceptable practice.

When I left the telecom industry and went into consulting, working remotely was the norm. If I was not at a client site, I was working from my home office. Having a presence at an "office" didn't make much sense, and it was a cost we could easily push out of the business.

But as I worked with more and more companies, I saw just how few companies were prepared to allow their employees to "work from home." I mentioned this to many clients over the years, but most chose to not create or prepare the infrastructure to support a remote workforce.

Then in March 2020, an unknown virus with origins in China, later known as COVID-19, struck the world. Offices were forced to close, and businesses were required to keep everyone out of the office. This made working a challenging task for companies that had never prepared for such an event. One such company still had the majority of employees in cubes, working from desktop PCs. The workforce had never embraced technology and most of the employees had no idea how to remote into their desktops. It was a real challenge.

For companies that had remote workforces, the business was able to limp along until the whole business was able to respond, and fully support remote work forces several weeks later.

The COVID-19 pandemic and the significant impacts the unforeseen pandemic had on the business environments world-wide, almost overnight should teach us that technology must be understood and leveraged to help businesses meet a rapidly changing world business environment. A digital transformation strategy must exist in every organization to support the business for just such unplanned and unforeseen catastrophic events. Businesses that can rapidly respond to new issues and challenges will continue to exist in uncertain times. Businesses that ignore technology and do not have a digital transformation strategy will face significant challenges.

CHAPTER 3

Understanding Digital Transformation Strategies

You've learned that it is critical to be as nimble as possible in our business, and the cornerstone of being nimble is the use of technology driven by a digital transformation solution. Having a clear and comprehensive digital transformation strategy is what makes implementing the digital transformation solution successful.

Most of what we do in our daily lives starts with a strategy. Let's use going to the grocery store as an example. First, I must think about what I need from the store. Because what I need drives which store I am going to choose. If I need eggs, then I will select my favorite grocery store down the street. If I need twenty bags of mulch for my yard, then I am going to select a big box home supply outlet, like Home Depot or Lowe's. My needs also drive what type of vehicle I will use. If I am going to get mulch, then I will need take my truck, and if I am going to get eggs then I can take my car or my truck - but with the truck getting 15 miles per gallon and my car getting 45, I'll probably take the car.

The point I am making here is that I often get called to client sites or invited to C-level discussions and the conversation starts with "I really need a new general ledger system" or "I want to save money." One of my first questions is, "What is your business strategy and how is it supported by your digital transformation strategy?" Like with the example above, the first set of decisions leads to a second subset of choices, like choosing a car that saves gas over a truck that may use lots of fuel.

As business professionals, we often see our business strategy de-coupled from our digital transformation strategy. But one cannot exist without the other. You may say to yourself after reading this statement, "Well I do have a business strategy but no, I don't have a digital transformation strategy." My response to that is, how will you execute your business strategy?

Often, I am told, "we need to be more profitable." Well, how do you know you are profitable now? How do you know if your profitability is above or below your peers? What is your measure of profitability? Mine is cash flow, or rather how much cash do I have left in the bank after I collect receipts and pay the bills. These questions will drive your solutions.

Let me share a quick case study to help articulate my point and lay a strategic foundation.

Case Study: Challenges that we face in business that require a solution.

Below are bullet points from notes I have taken during hundreds of interviews over the years. Think about the meaning of each point and how solutions for that point are normally approached. Then let's wrap them into a digital transformation strategy. Please note, these are not in any order of importance.

- Time zone
- Lack of quality
- Lack of training or knowledge
- Increasing demands for support
- Risk transparency
- Paper-based
- Workflow management
- Multiple languages
- Real-time status
- Inefficiencies
- Need for timeliness
- Manual processes
- Need for accuracy
- Lack of standardization
- Workload balancing
- Multiple currencies

For each Issue Statement in the table below, think about your response, what it means to you and how often you have struggled with each one of these matters.

Issue Statement	What does this mean to you? *(Fill in the blank)*	What I often learn after our discussion
Time zones are killing us		Departments are spread across the country requiring that one group works very early or very late.
We lack quality in our reporting		Reports are wrong. Spreadsheets contain errors or journal entries don't make it into offline reports
We lack training or knowledge		The resources I have do not possess the technical skills required in the new business environment

Issue Statement	What does this mean to you? *(Fill in the blank)*	What I often learn after our discussion
We have ever-increasing demands for support		The business is changing rapidly and accounting/finance is unable to provide information needed to make business decisions
Risk transparency		I have no idea where the risk really is in my business. I have audits, but what do they really tell me about my business?
We can't change because we are paper based		I'm two years from retirement and I really don't want anything to change for now, or this is my kingdom, go away
We have no workflow management		I have trouble closing my books because everyone does it their own way
We have multiple languages		Our teams lack the ability to collaborate over digital mediums
I need real-time status		I don't know how my business is doing until after the event has passed
We have inefficiencies		Operations, accounting, sales and marketing don't work together
Need for timeliness in our reporting		It takes way too long to get the books closed, or the information I get is no longer relevant
We have lots of manual processes		I am not sure how to drive inefficiencies out of my operational and reporting environment
Need for accuracy in reporting		I am not really sure what is the "right" number
We have a lack of standardization		In each area, we have key person dependencies that I have no way to overcome, and I am not sure what they do to get the job done
Workload balancing		My people work lots of hours, the newer employees are not willing to work 70 hours per week
We have multiple currencies		Foreign currency translations are killing me. We are not doing a good job of managing pricing or managing cash flow because of fluctuations

Let's jump to the common ask I get from my clients during the next part of the conversation. You guessed it, they want a solution.

Issue Statement	What I often learn after our discussion	Clients first suggested solution
Time zones are killing us	Departments are spread across the country requiring that one group works very early or very late.	I need a new GL system
We lack quality in our reporting	Reports are wrong. Spreadsheets contain errors or journal entries don't make it into offline reports	I need a new GL system
We lack training or knowledge	The resources I have do not possess the technical skills required in the new business environment	Maybe a new system can do the work
We have ever-increasing demands for support	The business is changing rapidly and accounting/finance is unable to provide information needed to make business decisions	I need to hire more people, but I don't have that in our budget. What system will fix this?
Risk transparency	I have no idea where the risk really is in my business. I have audits, but what do they really tell me about my business?	I need a system with a dashboard that can see across my business
We can't change because we are paper-based	I'm two years from retirement and I really don't want anything to change for now or this is my kingdom, go away	Can you give me a technology that solves my problem with disrupting my life
We have no workflow management	I have trouble closing my books because everyone does it their own way	There must be a system that I can install that does this for me
We have multiple languages	Our teams lack the ability to collaborate over digital mediums	There has to be a system that can do the work
I need real-time status	I don't know how my business is doing until after the event has passed	I need better data analytics
We have inefficiencies	Operations, accounting, sales and marketing don't work together	I need a new cloud based systems

Issue Statement	What I often learn after our discussion	Clients first suggested solution
Need for timeliness in our reporting	It takes way too long to get the books closed or the information I get is no longer relevant	I need a better integrated reporting and forecasting systems
We have lots of manual processes	I am not sure how to drive inefficiencies out of my operational and reporting environment	A new reporting system will solve my problem
Need for accuracy in reporting	I am not really sure what is the "right" number	What tool can I use to reconcile my books
We have a lack of standardization	In each area, we have key person dependencies that I have no way to overcome and I am not sure what they do to get the job done	Can I convert my close checklist into a task management software
Workload balancing	My people work lots of hours, the newer employees are not willing to work 70 hours per week	We need robotic process automation
We have multiple currencies	Foreign currency translations are killing me. We are not doing a good job of managing pricing or managing cash flow because of fluctuations	We need a new consolidation tool, ours is out dated, what will that cost?

The case study above illustrates that many business leaders focus on the issues in isolation. Usually, there is not a more global approach to the matter at hand. Each one of these solutions might have an adverse impact on the organization, when considered as a whole.

The illustration below provides insights into the common technology play:

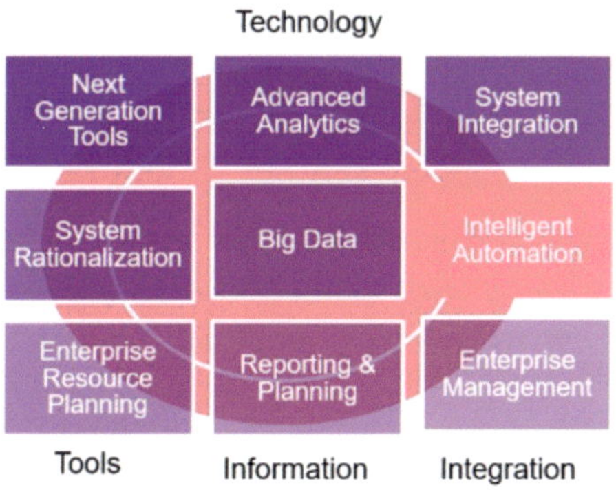

Considerations

Risks	• What can go wrong? • Organizational impact • Cost of control
Controls	• Placement of controls • People vs. Technology • Ability to audit
Governance	• What guidance to use • COSO 2013 vs. COBIT 5 • Do we have skill sets

Normally, this is what and how organizational leadership looks at issues. The focus is on either tools or getting information from a reporting system. Then the focus shifts to trying to integrate platforms across the enterprise so everything works together. Leadership usually has a secondary focus on risks, controls, and governance as they pertain to passing an audit - not necessarily on actual control and mitigation of the risk, just a tacit acknowledgement that it may or may not exist.

But What is Missing From the Illustration Above?

What is missing is the business need that drives each conversation, and prioritization for each set of activities or solutions to be deployed. Remember, an implementation without a strategy is most likely not going to meet the needs and objectives of the users.

So business needs and change management are missing from the conversation above.

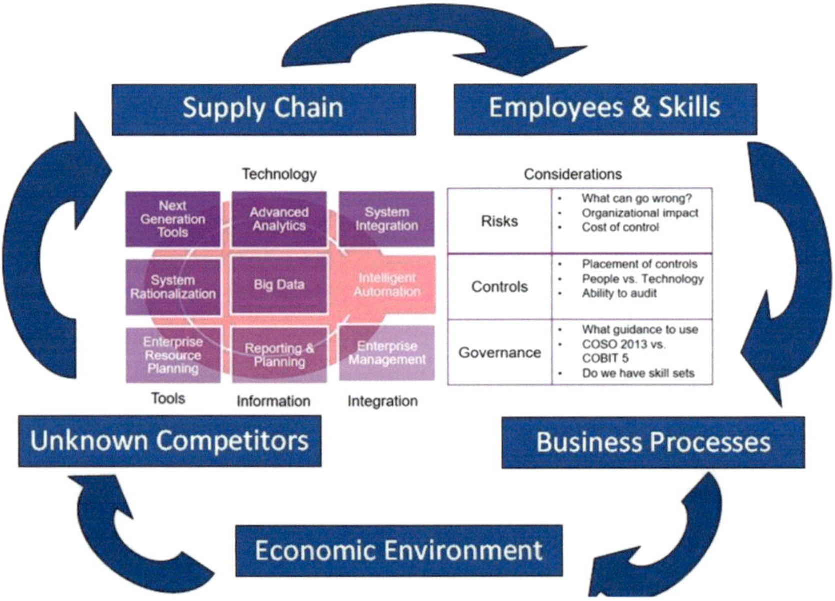

Like sharks circling your digital environment, the business is ready to consume your aspirations of digital transformation.

Each box illustrated above integrates in a four-dimensional matrix. Each one *must be considered in relation to all others* as you develop a digital transformation strategy. Intersection by intersection must be taken into account.

Intersection Analysis

Tools	Information	Integration	Business Environment	Considerations
Next Gen Tools	Analytics	System Integration	Supply Chain Employees & Skills Business Process Economic Environment Unknown Competitors	Risk
System Rationalization	Big Data	Intelligent Automation		Controls
ERP	Reporting & Planning	Enterprise Management		Governance

Intersections and considerations:

Next Gen Tools – Analytics – System Integration	Key Considerations
Supply Chain Employees & Skills Business Process Economic Environment Unknown Competitors	• What are the new tools that will best support my organization? *Example OneStream versus HFM* • Who in my team understands or can execute analytical review of our data • How hard will it be to integrate my systems? Cloud versus on-premise computing?

System Rationalization – Big Data – Intelligent Automation	Key Considerations
Supply Chain Employees & Skills Business Process Economic Environment Unknown Competitors	• Will the systems I select integrate with my supply chain data? • What can we do with the data we have at hand? • Can intelligent automation help me see competitors that are not obvious to the organization?

ERP – Reporting & Planning – Enterprise Management	Key Considerations
Supply Chain Employees & Skills Business Process Economic Environment Unknown Competitors	• Is my ERP the issue or is it *how it is being used*? • Do my employees have the training necessary to properly leverage the applications at hand? • Do reports meet my needs? • Can I forecast the future cash flow needs?

Risks - Controls - Governance	Key Considerations
Supply Chain Employees & Skills Business Process Economic Environment Unknown Competitors	• Is my data correct? • Does my data reflect my actual business? • Is my business prepared for any external cash flow shocks? • How long can we survive without strong cash flow? • What is my true competition? • Do my employees look around to see competitors that might exist in the form of changing technology?

We will not go through all possible integrations that could exist, but the information shows that there is a great deal to consider when looking at digital transformation and when developing a strategy.

I am guessing many of you who are reading this book don't recall a device called a "Palm Pilot" or "Paper Planner" which was a paper bound calendar that contained a place for notes, phone numbers and each day was broken into time increments. The "Palm Pilot" was an attempt to eliminate the paper planner version. The device would sometimes sync with your old PCs and could link to a calendar on the PC. But I must admit, I was one of very few people who could actually get sync to take place. But I digress. My point is that there are so many products that I used to use to manage my business and life in the past, that are now just gone.

Palm Pilot was a subsidiary of "US Robotics" and I would love to ask the leadership of that Palm Pilot subsidiary if they saw the market shift coming - and if they did, how did they prepare?

While doing research for this book, I ran across an article written in MIS Quarterly Executive, June 2016.

> *"Faced with the digital transformation challenge and the need to remain competitive in their industries, business leaders must formulate and execute digital strategies that drive better operational performance. Unfortunately, there are many recent examples of organizations that have been unable to keep pace with the new digital reality."*[3]

---

[3] MIS Quarterly Executive 15(2):123-139 · June 2016 [Page 2 of article]

The ability to implement a digital transformation strategy is and will continue to be a competitive advantage.

The cornerstone of the digital transformation strategy is preparation and a business focus. I will refer to the article above several more times, as I very much like how the authors frame their questions around a Digital Transformation Strategy. But what I want you to take away from this research and perspective is that the focus is mainly from an IT prospective. I am not suggesting that is wrong, merely that an IT focus removes the main driver, which is *doing business*.

Going back to the beginning of this book, we spoke about ledger paper, pencils and such. Recall your history. The invention of the lead pencil was a technological leap forward. The pencil resulted in users being able to "quickly" document business activities and such. But unlike today's technology, the pencil did not take on a life of its own. Always keep in mind that technology and IT is subordinate to the needs of your business. If it does not make things more profitable, then don't do it.

Digital Transformation Strategy

After all this research and discussion, we now get to the actual digital transformation strategy. But we must follow some simple rules.

Rule One

Always remember we are in business to make money.

Said simply, business is about cash flow. From non-profit to for-profit business, it is all about collecting and keeping as much cash in the business as possible. I know that many business leaders may say different things, but I have never met any entrepreneur who says, "I want to start a business and not make any money."

Rule Two

Digital transformation is to serve and support business.

It is not a pillar set by an IT team or any one group. It is about enablement, not roadblocks. The strategy must focus on how to help the business to be successful.

Just a side note on this rule...I recently worked with a mid-sized accounting firm. Overall, the business lacked strategy and IT appeared to run the business from a silo. I never really knew what goals were set and by whom.

The IT organization made it nearly impossible to get anything done. When they were questioned about why actions were being taken, they would respond with "it is a security issue." The IT team was bloated, expensive and unresponsive to business needs. I am not blaming the IT organization because it was the overall business that lacked an integrated digital transformation strategy, driven from the top. As accountants, we are always told to analyze risk based on the "tone at the top," but it is often ignored when it comes to digital transformation strategy. Many years ago, by a great mentor of mine at the large telecommunications company where I worked told me that "nature abhors a vacuum. If one exists, it will get filled." In this case the vacuum was a true lack of strategy and leadership. So the IT organization filled in the void – it is just that simple.

Rule Three

Digital transformation strategies can't take years to help the business.

A digital transformation strategy can look forward, one, three, five years and beyond. But the strategy can't be for the sake of strategy alone. A digital transformation strategy must be measurable and attainable in small, medium or large bits. It must help today, tomorrow and in the future.

When Y2K was discussed in 1999, there was a great deal of consternation about what was going to happen. From the lights shutting down, to cars that would no longer start because the on-board computers would not function past 12/31/99 because the PC would believe it was 1900. Or so the fear mongers told us. But the clock was ticking, pun intended. Organizations had to quickly develop strategies, and build and test them before midnight December 31, 1999, lest we all turn into pumpkins. Good news is that it turned out to be an incredibly expensive drill, maybe for not. But the point is that we had a known event that was going to hit on a specific date and time. Most of what we face in business is not that clear cut. Take the COVID-19 scare of 2020. It came upon the world stage in just weeks and paralyzed over 150 nations. (History will tell if it was really an issue that required destroying businesses and lives to save lives, or not-that is for another book). That is why digital transformation strategies must be actionable now, so that business can always be at least somewhat prepared for the unknown.

Rule Four

Digital transformation strategies must be communicated to the organizations, with owners and accountability clearly defined.

Digital transformation strategies require timelines, milestones, deliverables, budget commitments and ownership.

As an example of the "why"...I worked at a large retail merchandiser. The CFO was a very competent accountant but a very poor technology strategist. This CFO had a great deal of trouble getting ahead of the business needs and spent most of his time just reacting to what he believed to be the most pressing needs. Very little thought was given to how the technology was being deployed. Even less thought was given to root causes, and when we provided thoughts, guidance and roadmaps they were ignored.

The CFO complained that the projects had not "gone" well and were behind schedule, but not once did he ask what could be done differently, so that technology could be better used to support the business. No questions were ever raised about what digital transformation strategies should be applied or what would be the next steps. Risks around internal controls over financial reporting and IT general controls were mostly ignored because in the end, there was little if any communication within the organization about how people, processes and technology would fit together and how lack of integration could result in business risks.

To put a framework around the issues noted above, I want to use the information contained in the article "MIS Quarterly Executive, June 2016." A table they developed does a good job of clarifying key "Decisions" that will support the development of a "Digital Transformation Strategy." Although this table is focused on the "media" industry, you will notice it applies to other industries as well:

Table 6. Key Decisions for a Digital Transformation Strategy[4]

Use of technologies					
1. Strategic role of IT?	Enabler	Supporter			
2. Technological ambition?	Innovator	Early adopter	Follower		
Changes in value creation					
3. Degree of digital diversification?	Electronic sales channels	Cross-media	Enriched-media	Content platforms	Extended business
4. Revenue creation?	Paid content	Freemium	Advertising	Complementary products	
5. Future main business scope?	Content creation	Content aggregation	Content distribution	Management of content platforms	Other
Structural changes					
6. Responsibility for digital transformation strategy?	Group CEO	CEO of business unit	Group CDO	Group CIO	
7. Organizational positioning of new activities?	Integrated	Separated			
8. Focus of operational changes?	Products and services	Business processes	Skills		
9. Building of competencies?	Internally	Partnerships	Company takeovers	External sourcing	
Financial aspects					
10. Financial pressure on current core business?	Low	Medium	High		
11. Financing of new activities?	Internal	External			

As you review the table, you see that IT is an "enabler" and "supporter" but not a driver or a true "decision" maker. You can also see that there is a structured approach to a digital transformation strategy. There are probably hundreds or even thousands to choose from, but I like the simplicity of the one above.

Before we get to development, we need to dissect the intersections shown above. The authors of the article have broken the Digital Transformation Strategy (DTS) into three buckets. I will break these buckets down into a more generic format that can be applied across various industries.

As the author states: "Although this analysis is based on the media industry, we believe that apart from the questions relating to a firm's value creation, the findings can be transferred to other customer-oriented industries. The value creation dimension usually varies significantly across industries and business models."[5] You might say to yourself, well, this does not apply to me, because I am not a "customer-oriented" business. I would say you're missing the point of your business because *without customers, you have no business*.

[4] MIS Quarterly Executive 15(2):123-139 · June 2016 [Page 35, Table 6]
[5] ibid

Consider the table above and the general theme below:

Bucket	Business Driver	Consideration
Use of Technologies	Generate revenue or reduce expense, or both	How far or fast do we go as an organization? Does fear of risk make us a "follower?"
Changes in Value Creation	Where are we going to generate revenue and what does the future of our business look like	What sales or revenue generating channels do we have? How do we get better in those channels? What future opportunity are we not seeing in our business?
Structural changes	Is our organization capable of driving and supporting a DTS?	Have the right leaders been assigned roles and responsibilities to help support DTS change? Will DTS be integrated or separated into silos to show quick wins?
Financial Aspects	Do we have the funds in the business to implement a DTS? Will a DTS put additional financial pressure on current core business? How can we finance DTS?	Can our business afford to implement a DTS? If your business is unable to implement a DTS, then how do you grow or maintain your current business position? Where to get money? Raise capital, borrow, cut costs in various areas of the business?

All the considerations noted above must be discussed in an open forum with all business leadership. Without such discussion, DTS becomes nearly impossible to successfully execute.

I often hear that technology or digital transformation strategies "don't apply to my business or industry." A research paper written in 2015, *Strategy, Not Technology, Drives Digital Transformation: Becoming a Digitally Mature Enterprise,* shows that digital transformation is not driven by industry, but is a part of business in general. It also explains why, over the years, the industry-specific questions and asks of me regarding industry have largely subsided.

As the article illustrates, it is a strategy that drives digital transformation, not technology. Since digital strategies apply to all industries, there is no longer a need to look at a business from only the industry prospective. Industries can change very rapidly. We can look to our 2020 example, in which Kindergarten through high school students were ordered to stay home because of the COVID-19 pandemic. Overnight, distance and electronic learning or video conferences became the norm. Who would have predicted that event?

By nature, some industries are ahead of others, but as the author suggests in his title, "*Industry Lens: Leaders, But No Laggards*."[6]

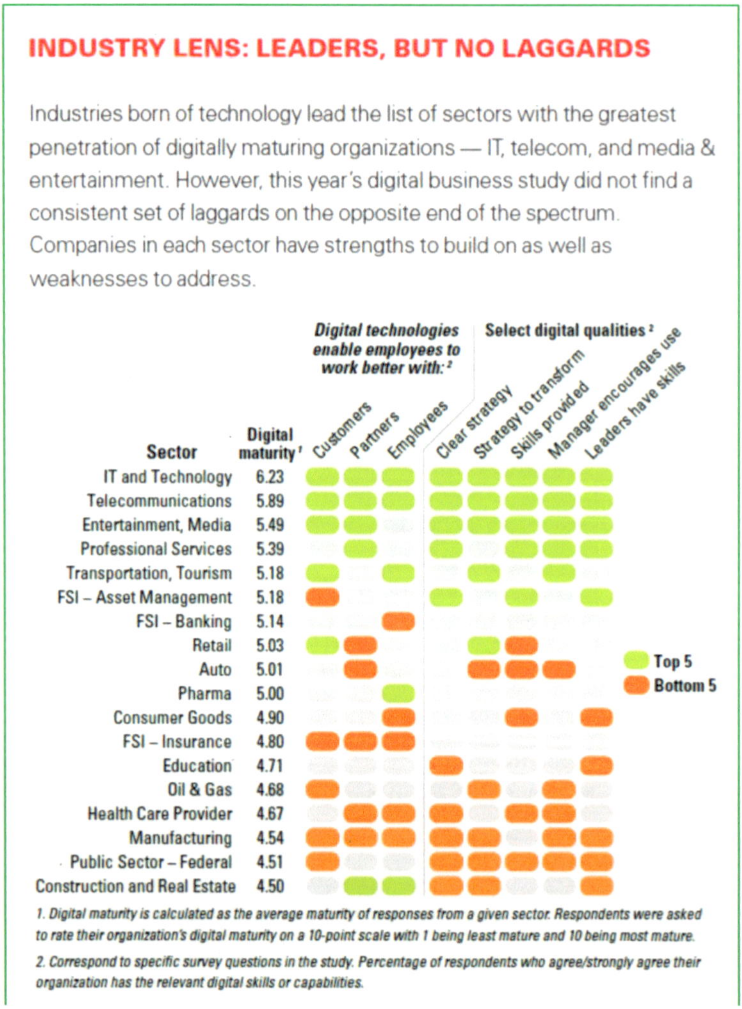

Clearly, EVERY industry must consider a DTS.

---

[6] Strategy, not Technology, Drives Digital Transformation: Becoming a digitally mature enterprise SUMMER 2015 RESEARCH REPORT By Gerald C. Kane, Doug Palmer, Anh Nguyen Phillips, David Kiron and Natasha Buckley, page 9

Building a Digital Transformation Strategy

It is *strategy* - not technology - that should drive your DTS. The authors of "*Strategy, not Technology, Drives Digital Transformation: Becoming a digitally mature enterprise*" described their approach clearly in their statement:

> *"Imagine an ideal organization transformed by digital technologies and capabilities that improve processes, engage talent across the organization and drive new value-generating business models."* [7]

Let's take this a bit further to a mini case study discussion.

Issue: A mid-size company has trouble closing their books in an efficient and timely manner.

Effort: A great deal of effort and manpower is expended to close the books.

Accuracy: Reports are not very helpful and often late or contain errors.

I recall sitting in a room with a few Big 4 partners and other consultants, when in walks a CFO. He is a very direct person who knows his business and wants to focus his attention on the business, and not on reporting.

The partners start off asking about the issues and possible solutions. The CFO sits in his chair and says, "Why can't I just push one button and all my books are closed?" The room erupts in laughter at the statement. The CFO, upon hearing the response, excused himself and we left. No work was won that day. But I didn't laugh at the CFO's response, because as a practicing accountant, I had asked that same question a million times at 2:00 AM on a Saturday morning, waiting from my sub-ledger to clear so I could finally post and go home and get some sleep.

Goal:

The goal of any strategy should be driven towards what the CFO in this case study wanted - an efficient way to close his books, without taking up everyone's time and taking their focus away from the business.

---

[7] Ibid, page 5

Building Blocks:

Rules are just that, unless they can translate into meaningful action. When you are asked by your CFO to do something to "make us more profitable," what does that really mean? Now, contrast that with another rule. It is illegal to drive your car through an intersection if the light facing you is red, and you will be fined or injured if you choose to violate that rule. Your DTS should be just as concrete and understandable.

Below is an example of putting this into action - how I take ideas and drive leadership into taking action, all while forcing them the keep the overall DTS in mind:

Rule	Strategy needs to address	Strategic Action
Always remember we are in business to make money	Reduce external audit fees by 50% in two years, so that we can save money and reduce demands on our staff	Poll the teams and ask about what is being audited. Confirm those areas are important to managing business risk. Ask the team for suggestions on how to reduce errors in systems that lead to audit findings Analyze if it is a lack of training, system integration or legacy system that is driving the large amount of audit hours
Digital transformation is to serve and support business	We need to reduce the amount of effort it takes to apply cash to outstanding invoices	Review processes Review skill sets Review management oversight Confirm if issue lies with people, process or the technology that is being used Take action that supports DTS goals
DTS must be communicated to the organizations with owners and accountability clearly defined	IT is a chokepoint for our organization and is seen as an impediment to the business.	C-level leadership to define IT role Bring in outside expert to confirm IT role is appropriate for you firm Work with IT leadership to ensure they understand they are there to support business, not be the business Focus IT on tasks that enablement

Rule	Strategy needs to address	Strategic Action
Digital transformation strategies can't take years to help the business	Our DTS will incorporate a new Point of Sale system integration with the general ledger so we can better understand cash flow on a daily basis	Ask Point of sale teams what is working and not working Ask managers in the Point of sale team if they used processes or software at other jobs that worked exceptionally well Ask the reporting team if all data elements are available for them to use sales data in the General ledger system Speak with peers about what and how they approach the same issue Attend conferences that focus on your space

Now you, the reader, can fill in the blanks for yourself. Pick an area of your business that you need to address and complete the following chart.

Rule	Strategy needs to address	Strategic Action
Always remember we are in business to make money		
Digital transformation is to serve and support business		
Digital transformation Strategies must be communicated to the organizations with owners and accountability clearly defined		
Digital transformation Strategies can't take years to help the business		

It is my hope that you consider building your DTS with/as both a top-down and a bottom-up approach/implementation. Just starting at the top, without bottom actions will most likely never work. Likewise, just focusing on one issue or aspect of the business will most likely only yield small or silo-specific benefits to the business. This will give those in the business who do not want to change, the opportunity to say "see, I told you it wouldn't work."

Building your Strategy

While I do not believe there is only one way to approach a DTS, in Table 6 below from MIS Quarterly Executive, June 2016, the authors lay a strong foundation for what decisions need to be made.

Table 6. Key Decisions for a Digital Transformation Strategy[8]

Use of technologies					
1. Strategic role of IT?	Enabler	Supporter			
2. Technological ambition?	Innovator	Early adopter	Follower		
Changes in value creation					
3. Degree of digital diversification?	Electronic sales channels	Cross-media	Enriched-media	Content platforms	Extended business
4. Revenue creation?	Paid content	Freemium	Advertising	Complementary products	
5. Future main business scope?	Content creation	Content aggregation	Content distribution	Management of content platforms	Other
Structural changes					
6. Responsibility for digital transformation strategy?	Group CEO	CEO of business unit	Group CDO	Group CIO	
7. Organizational positioning of new activities?	Integrated	Separated			
8. Focus of operational changes?	Products and services	Business processes	Skills		
9. Building of competencies?	Internally	Partnerships	Company takeovers	External sourcing	
Financial aspects					
10. Financial pressure on current core business?	Low	Medium	High		
11. Financing of new activities?	Internal	External			

[8] MIS Quarterly Executive 15(2):123-139 · June 2016 [Page 35, Table 6]

To help you consider how to put it into practice, I have converted Table 6 above into a generic building block framework:

How will Technologies support my business via DTS

1	What Role will IT play?	Enabler			Supporter		
2	What Should our goal be for using Technology?	Innovator		Early Adopter		Follower	
2.1	What is our technology fear factor?	Too much risk?		Reduce risk to business?		Lots of business risk?	
Using DTS to create organizational value							
3	How much technology will we use to support our business?	Accounting & Rapt - Process automation	Operations	Accounts receivable	Supply chain	Application integration	Security
4	Generate revenue or reduce expense	Reduce labor expense	Data analytics	Automation		End-to-end integration	
5	How to support or see future business	Analytics - Big Data - Data Lake					
Need to Change Organization Structure							
6	Responsible for tone at the top, digital transformation Strategy	Owners	CEO	Operations Owner		IT	
7	How to execute	Integrated			Separated		
8	Change management	What we do to make money		Processes		Build Skills	
9	Build DTS competency	Internal		Consultants		3rd party	
Cash Flow Impacts							
10	Setting the budget	Year One		Year Two		Year Three	
11	Source of Cash	Internal		New Business		External	

Building Blocks

Like a toy LEGO®, each part of a Digital Transformation Strategy will fit together with any other part, based on your imagination. However, it might just look like a wall with several different colors. It's best to build with a vision and a goal in sight.

How to use the building blocks:

How will technology support my business?	Consideration Area	What works
What role will IT play?	Enabler/Supporter	The business must define the role of IT. Don't let IT drive the business. They are usually not customer facing and all too often focus inward and not outward.
	Goal for using technology	Being a follower is not a good strategy. Those who follow are usually at the end of the line. So either focus on early adoption or innovation. This will depend on your ability to change and manage your business at the same time
	Your perception or "appetite" for using technology	Most of the time, when I speak with business leaders, they fear technology. Either costs, or the fear that the spend will not create the value, scare them. They fear that the change will be too disruptive or just not work because it "sounds" too good to be true. Focusing on the risk of doing nothing often gets people to think about the future.
Using DTS to create organizational value	**Consideration Area**	**What works**
	Where and how to use technology to support business	Don't just look at one area of the business. Look across the whole landscape. From order to cash and all things business. This is not an accounting fix or a supply chain fix. As one of my mentors always said, you must look at all of it.
	Generate revenue or reduce expense	You are in the business of making money. So focus on generating revenues or reducing expense. My father used to tell me, "it is easier to save a buck than it is to make one." Don't be short-sighted, but don't lose sight of the goal - money.
	How to support future business	Focus on how technology can help you to see into and prepare for the future. Don't just focus on the words, like "big data," "analytics" or "automation." Think about what that means to your business.

Need to change org. structure	Consideration Area	What works
	Assigning responsibility	Assign the right responsibilities and authorities to your team. Don't try to manage it from the top and don't leave all choices to a lower level manager.
	Execution	Identify and agree upon milestones, deliverables and timelines. Make sure that this is communicated before any actions are ever taken.
	Managing change	Communicate, Communicate, communicate – and collaborate. Do not work in silos.

Cash Flow	Consideration Area	What works
	DTS budget	Digital transformation can't happen without a budget. Make sure you have a plan. Look at year one, year two and finally year three. Given how rapidly change occurs .
	Finding the dollars	• Budget dollars can come from moving dollars from one department to another • Shaving expense and using these dollars to fund initiatives. • Dollars could also come from building out new lines of business or sales to fund changes. Organizations also obtain funding from banks, investors. *The bottom line is the bottom line. DTS must be funded to be successful.*
	Building your competency	Communicate honestly with the team. Some people can manage and some can do. Let the team fairly and honestly provide feedback. The truth must be told. In Good to Great, Jim Collins calls it "getting the right people on the bus and the wrong people off the bus."

At this point, the DTS foundation has been laid and can now be placed into action. Every business organization will look a bit different, but the building blocks should always be the same. Focus on your business needs, short- and long-term goals, and continue to consider competitors as you build out your digital transformation strategy.

CHAPTER 4

A Practical Approach to Applying Digital Transformation Strategies

The past chapters focused on how to prepare and design your DTS. Now we will move from the theoretical to a practical approach to executing the DTS. To lay the foundation, I'll take you through a few case studies.

Case Study – No Strategy

I worked with a mid-sized electronics sales company that was experiencing issues with their tax reporting, their general ledger balances, and the accounts receivable process. I was asked to come in and help the organization address each issue. From the start, I asked the CFO if they had any digital transformation strategies or if they had an IT strategic planning initiatives. I wanted to know if the company had any plans to work with IT to help identify any data or data transformation issues. I also asked if there were any committees that were focused on process improvements or if route causes had been discussed.

The CFO said that no centralized strategy existed, and no official budget had been allocated to solve the problem. The CFO also had a very limited understanding of what actually was causing the problems for her organization.

The CFO went on to say that they were having issues with their external accounting support staff. For this leader, the goal was to get things fixed right away, at the lowest cost possible. I had been doing this type of work for years and saw this as a major project risk.

Since the CFO had no idea what was on the horizon with the IT organization (and if she knew, it was not being shared), this meant that any solution provided or executed would take place in a silo and most likely not address any root cause. The result of executing a project in a silo would be a perception of project failure, and that perception of project failure is a major project risk.

The CFO also never formally assigned anyone as a project leader on the client side and did not make anyone on the client side accountable. My advice to review the current state, identify issues, possible solutions and institute a steering committee to assign roles and responsibility were ignored by the CFO. However, the team I was working with wanted me to proceed with the project work, so I did.

Roll forward several months, and the applications requested were installed. To the client's surprise, the applications didn't produce as planned. The main cause was not a technology issue, but a data issue. The organization lacked any real controls over operations, reporting or master data management. The data was incomplete, inaccurate and was not timely (data crossed periods).

Since we had no overall transformation strategy, the CFO was not able to affect change in controls over data. Additional manual processes had to be included with the new software, just to get things running. This increased the time to complete the project and degraded the value the project and the technology could have produced.

Case Study – Strategic Framework

My next case study is focused on a client that produced pharmaceutical equipment. I was called in to address a control issue related to the monthly close. The client had brought in a new controller to oversee a transition in the business. The new controller found that many, if not all key general ledger accounts had either never been reconciled or had not been reconciled in several months.

The controller made me aware of a larger ERP implementation project that was being led by a Big 4 accounting firm and advised me that his need would be part of an overall transformation strategy.

The controller assigned me two accounts to support the effort, identified an IT support team member and together, using a project charter we assigned roles, responsibilities, milestones, objectives and success factors.

This project was viewed by the company as a huge success. The application was up and running as per our project plan, client accountants took over the processes and learned the application. The team also integrated with the large transformation initiative and the application was quickly and efficiently integrated with a large ERP platform some months later.

As the character Hannibal Lector (played by George Papard) used to say in the 80s television show *The A-Team*, "I love it when a plan comes together."

This client not only had a project plan, but also included their efforts into an enterprise-wide digital transformation effort.

Case Study – Foundational Strategy

While working on this book, I had a unique opportunity to discuss the effects of a world-wide pandemic on an oil and gas supply company. The company sells and services boring equipment and other heavy equipment used in farming. It has retail outlets and over 345 employees, spread out over the central part of the United States.

Starting several months before the pandemic struck, the CFO of the company started to look at how technology could help his business to be profitable and to grow both organically and via acquisition. The other goal was to drive IT costs out of the business and to provide better reporting and business analytics.

As a result of this effort, the company implemented business intelligence software and migrated to a new point-of-sale system. The company struggled to identify an IT leader but moved forward anyway.

When the pandemic struck, businesses were forced to shut down operations, which resulted in catastrophic business losses. This downturn in business had an impact on drilling and heavy equipment sales. But to the surprise of the CFO, revenue remained stable. When the CFO reviewed the data, he found that sales of new equipment were down, but sales of service and replacement parts increased. This shift made sense to the CFO and his team, who had been at the company for over 30 years each and observed significant business shifts over the years.

Result of Hybrid Digital Transformation Strategy

After my discussion with the CFO, I asked, "What was the greatest benefit from the company's decision to start upgrading reporting and sales systems?" He replied that the greatest benefit was that he knew the data was accurate and timely. This knowledge eliminated the need for his accounting team to waste time on verifying the data. The data met their assumptions and the C-level team was able to quickly respond to the almost instantaneous shift in the business.

The C-level team also supported issuing every employee a laptop, so when all employees were forced to work remotely, that was possible. Infrastructure had been upgraded and a Director of IT had recently been hired. The ability for everyone to work remotely produced a significant value to the business. According to the CFO, "my whole perspective on working remote has changed. Prior to the COVID (global pandemic) I did not support working remote. But what I saw was the same or better response from my accounting staff and an increase in overall productivity."

This case illustrates how laying down at least some type of digital strategy helped this manufacturing business not only to meet the challenge it faced, but to overcome the challenge and keep the business running at nearly the same level of profitability.

Applying the Digital Transformation Strategy

The best approach to applying a DTS is to start small.

First, what is your DTS? What are the first areas to be addressed? What is the organization's budget and what approvals or decisions need to be made before you start?

Based on the statement above, let's assume that the DTS has already been established and is ready to be executed.

DTS Steps	Leading practice	How I have observed this work
Understand the issue facing your organization/team	• Speak with your team • Understand their pain points • Confirm organization needs versus effort required to support business needs • Speak with other business units to see if they have the same issues	C-level leadership often believes there are few issues. Most have little understanding of the effort exerted to close the books or maintain and grow the business. This is not because the C-level leaders don't get it. This is often because we as the provider of information don't communicate how much work it takes to get things done for our leadership. Once leadership understands the effort to meet their needs, it is time have team meetings and "white board" or processes flow sessions. These are an excellent way of identifying issues and getting everyone in the organization on the same page.

DTS Steps	Leading practice	How I have observed this work
Determine where your issue fits within your organizations DTS	• Look at your organization and understand how this fits into your organization's DTS • Understand how many parts of the DTS may be supported or delayed due to addressing this issue • Understand how your issue will help make your company more money or help save money	Each section of the DTS should be reviewed and compared against the other sections. Competing interests are always present, so logic and business impacts must be used to guide action. Usually, this part of the process will result in some horse trading to get things moving forward. But organizations that focus on DTS and business alignments quickly find common ground.
Prioritize your issue against the DTS	• Set a priority list so that you know when, where and how you will execute the plan • Weigh priorities against the unknown • Communicate the priorities	After organizations decide what bucket the issue sits in, they then can move to prioritizing which effort will be initiated. I have often observed two or three initiatives boil to the top in this instance. Often they are executed at the same time as long as the organization can support the change.

DTS Steps	Leading practice	How I have observed this work
Explore solutions	• Understand where the solution might lie. It most likely will cover many areas • People – Do I have the right people and skill sets to lead or manage the solution? • Process – Are the processes part of my solution or creating the problem? • Technology – Will technology solve my issue or cause more problems? • Cost benefit consideration on consuming budget	Exploring the different solutions ends up being the most time consuming effort. Most organizations end up being paralyzed at this stage without a strong adherence to the topics noted above. If competing interests, budgets and personalities have not aligned, then this is where your solution will result in failure. The most successful organizations adapt an agile approach to the change. They look at what will work to meet most of their needs in the one year time horizon. This way they team is able to make a positive change in the business and show a quick win. Gone are the days of the multi-year projects that take so long to complete, the people who finally see it come to life, don't even recall why it was initiated.

DTS Steps	Leading practice	How I have observed this work
Identify solutions	• Identify solutions that best support business and DTS needs	Organizations that focus on the quick wins, responding to their environment and moving forward are the most successful at identifying a solution that works for them. Organizations that never fully reviewed DTS setup efforts usually end the efforts here, or a strong influencer in the organization drives to get a solution that meets their needs. This is not necessarily all negative, but ultimately may end up costing the company more money because near future changes in IT or business environments may require changes to the solution that was chosen.
Implement solutions	• Execute using a rapid or more agile approach • Set tasks, milestones and deliverables • Identify roles • Identify responsibilities • Leverage project charters • Develop and use a project plan	Implementation comes down to good old fashioned project management. Project management has been around for a long time, but is often ignored. Unfortunately, most organizations I have observed leave project management at the end of the train. Leaders tend to see this as a non-value added set of activities. But like any great symphony, it is usually great because it has a great conductor. The most successful organization execute based on or against a plan. The organizations that take a methodical, managed and thoughtful approach, starting with a project charter and ending with a living project plan end up with the most successful outcomes and generate the greatest value for their organizations.

Ask yourself, how would you or your organization normally approach DTS? Also, consider your last software purchase, process change or response to external pressure. Was it based on a strategic thought process, or was it just reactionary?

CHAPTER 5

Technological Solutions: Microsoft Excel to Intelligent Automation

In the prior chapters, the points were identifying business needs and how to align those with a viable digital transformation strategy.

In this chapter, we will cover multiple system and reporting technologies, from those that have been around for several years to the more current, intelligent or purpose-built automation systems. We will cover how to implement those technologies and how they are used, while working within or considering your DTS.

Excel, for example, is a very useful tool. But all too often I meet with clients who say, "Oh, Bill has us all automated." After a few moments of questions, I find that "automation" is a macro or checklist. Now there is no right or wrong use of Excel, but the misconception that macros are automation will not help your organization implement an effective DTS.

Below is an overview of Excel:

Pros	Cons	Experience
• Easy to use • Standard part of MS 365 professional • Taught in all business schools • Macro enabled • Formula enabled • Multiple sheets can be used to build sophisticated models	• Easy to use* • Visual basic macros are not easy to build, maintain or troubleshoot • Formulas are easily corrupted • Only a small amount of data (1M) records can be loaded • People-centric	• Excel is 1990's technology and is suited for smaller data sets. • Producing dashboards, reports and what-if analysis takes time and often more expertise than most organizations possess.

*Note: Something that is easy to use can be good, but there can also be a downside. The downside is that users don't want to abandon what they feel works for them and their team. The simple technology becomes an extension of the person's role and often results in a key person dependency. In the accounting industry, we call this the "mother of all spreadsheets."

How it is implemented – Excel is the cornerstone of most, if not all, business reporting and management. It is used from the smallest organizations to the very largest. But unfortunately, Excel lacks control, integration and version control, just to name a few. Over the years, various applications began to take shape, such as SharePoint or other collaborative software applications to support collaboration, but Excel remains a user focused point solution. A point solution is one built by a user to just meet their job needs. It does not scale or support enterprise-wide use.

So at best, it is a point solution and used by various teams and skill sets. In many ways, Excel is our 10-key tool of the 2020s and beyond.

There is no real formal process to use Excel or integrate it into any possible DTS that you may have executed - Stay this course at your own risk.

Relational Database

I won't spend too much time on relational databases, although it is the most common form of business systems in use today. These systems date from the 1980s and before. Older companies may have mainframe computers or what was called "mini" computers, like IBM AS400 systems. These systems are very reliable, stable and secure. But that also makes them very inflexible and hard to use. When I encounter these older systems, the goal is usually to migrate towards a more modern system. Many point-of-sale systems and supply chain systems are managed by such devices, and are often called by the information technology teams as "legacy" on-premise equipment.

Migrating from Legacy Mainframe

A successful migration will require that the solution is included as a key part of your DTS.

Aligning solution with strategy:

DTS Steps	Approach	Implementation the Change
Understand the issue facing your organization/team	• How do you know that the application is causing your issue? • How do you know a new system will make things better?	• Form a project team • Collect information from stakeholders across the organization regarding the change • Determine the criteria for success
Determine where your issue fits within your organization's DTS	• Where does the business issue fit? • Is the issue isolated to one segment of the business?	• Project team will need to understand the issue across the people, the business processes and the technology

DTS Steps	Approach	Implementation the Change
Prioritize your issue against the DTS	• Prioritize issues and include as a key part of the DTS and priorities	• Produce a gap analysis report to compare issues against DTS to help drive the project metrics
Explore solutions	• Understand what each solution will offer and cost of implementation • Look at solution in terms of disruption to the organization	• Have the team compare different solutions • Include/weave solution into formal DTS project timeline and plan
Implement solutions	• Set the solution budget • Assign roles and responsibilities Define and confirm with leaders what is considered "success"	• Identify owners in each stakeholder group Set milestones, tasks and deliverables Try to plan for unexpected challenges Confirm how conflict will be managed Apply an agile approach

Migrating from a legacy system to a more modern application is a very mechanical process. For instance, if you are going to migrate from QuickBooks to NetSuite, the approach above is one that would work very well. Even if your organization is small, it is best not to just jump into a technology without considering your DTS. Technology may be the last play. People and processes may be your more pressing issue.

Intelligent Automation

Migrating to an intelligent automation will take a more focused approach. But before we start, it's critical to understand the role of success factors. I have spoken with many friends and colleagues in this space. Some say that "intelligent automation" does not work because the benefits are never realized. The thought is that costs are too high and benefits are too low.

Let's use a case study to dive a little deeper into this thought process.

Case Study

A very small shop wants to automate, using robotic process automation software.

Issue: It will cost anywhere from $5,000 to $25,000 to purchase a "BOT" license and at those prices, I can't reap an economic benefit because I can't replace "one person."

Scenario One

If the goal is to remove one complete headcount, then taking small parts of a few jobs will make this process more of a challenge. But let's turn this around and take my advice above and go not from a technology perspective, but from a time-in-motion study perspective.

Let's say you have four (4) employees in your group and for the sake of argument, they are all in your accounting team. One is focused on accounts receivables, one is focused on accounts payable, one is focused on inventory and supplies, one is focused on reporting and paying taxes...and then you have a manager. Under normal conditions, if a resource leaves, you would work to replace that resource. But what if there is a world-wide pandemic and you have to reduce your headcount by two employees. You, being the prudent company leader, take the newest member of the team and perhaps the highest cost individual and you send them into furlough purgatory.

There are only three accountants remaining and your manager. Do you say to yourself, oh, I can't let these people go, because I can't fully transfer the work or change "any" processes? The answer is a simple "NO" because you have to eliminate the headcount to remain in business.

What does one do? We do what we have always done. Look at our people and see who has the skill to do what jobs. Everyone pitches in and takes over different parts of the work and we change our processes to absorb the extra work. If we are lucky, we all stay employed, overcome the hurdle and make it to the end of the crisis. And the work goes on and continues to get done.

Thought Challenge One

As I present the case study above, none of you reading this will say, "oh, the author is crazy." Why? Because we have all been through these drills in the past. Now, the last world-wide crisis was unique, but the issue we faced in this crisis is not - doing more with less, under lots of scrutiny and pressure.

Scenario Two

Same scenario as above, except instead of a crisis, the business makes a choice to leverage Robotic Process Automation (RPA) or similar intelligent automation technology to eliminate the effort and drive cost out of the business.

Thought Challenge Two

What I normally get as we approach the same topic above is that "we just can't do this effectively" or "it is just too expensive to make it happen." But consider the same approach. What could your team collapse into a single position or eliminate if required?

Findings

In this case, I hope that it is clear the approach will occur at the micro level. The business process owners will need to manage changes to their own process, just as it would if they were forced to reduce headcount. Consolidate, eliminate and change if necessary. So we can start to see how RPA-type technology could be used to eliminate effort.

In all honesty, medium sized and very large companies take the same approach as above, but they are able to dedicate larger teams and focus to such efforts. This results in a greater success and the perception that RPA or intelligent automation only works at large companies, but that is just not true.

Implementing RPA/Intelligent Automation

When it comes to automation, the same rules apply; look at your DTS and confirm where the technology fits into strategy. Once that has been confirmed, then move towards the build stage. To move into the build or implementation stage, we start with a process approach that I have developed and used over the last five years. NOTE: Due to the large amount of information related to RPA and software BOTS I won't take the time to rehash the past. If you're interested in learning more about RPA, please visit my series on the topic, located at the AICPA Competency and Learning Center https://competency.aicpa.org/media_resources/212520-implementing-robotics-process-automation-rpa-appl.

I have outlined some key steps below:

- Identify key steps in implementing an application/process change
- Recognize key considerations in deploying the change (Think DTS)
- Integrate approaches to implement into your DTS to support business objectives.

We can walk through this at a high level.

Identify Key Steps in Implementing an Application

First, consider what type of software BOTS you want to implement. Software BOTS fall into two main categories. - centralized and decentralized BOTS. This means that the software BOTS are managed from a main location or group (centralized), versus software being loaded onto various laptops (decentralized). I won't get into the control discussion in this book, but be on the lookout for papers related to the subject of controls around software BOTS and discussions related to Information Technology General Controls.

Once the decision is made toward centralized or decentralized, the applications can be selected and loaded onto a Personal Computer or various Personal Computers.

DTS and Key Considerations in Deploying the Change

After the software is procured, the organization must choose a deployment strategy. There are two major approaches when deploying BOT software - an "agile" approach or a traditional "software development" approach.

I will provide you with a bit more information about the two options and how to approach each with your DTS.

Agile	Software Deployment
Build the BOT Test the BOT Deploy the BOT This is at a very granular level, with a goal that when all the relevant "steps" are complete, the entire process selected is automated.	Identify processes Develop functional requirements Build and code out software Test software Deploy the software In this approach, it takes many months, if not years to complete and automate an end-to-end process.

The software development approach often takes much longer to produce, mostly because requirements are gathered, code written, then tested and finally executed. This process is very formal and takes a great deal of coordination and organizational effort.

On the next two pages are detailed examples of each of these deployment approaches.

Agile Approach

How will Technologies support my business via DTS							
1	What Role will IT play?	Enabler		Supporter			IT will set up machines and load BOT software
2	What Should our goal be for using Technology?	Innovator		Early Adopter		Follower	Show quick wins for organization
2.1	What is our technology fear factor?	Too much risk?		Reduce risk to business?		Lots of business risk?	Risk averse = Centralized Risk tolerant = Decentralized
Using DTS to create organizational value							
3	How much technology will we use to support our business?	Accounting & Rapt - Process automation	Operations	Accounts receivable	Supply chain	Application integration/Security	Focus on one area of the business, build and hove/show a success in a very short time
4	Generate revenue or reduce expense	Reduce labor expense	Data analytics	Automation		End-to-end integration	Align BOT build with business needs
5	How to support or see future business	Analytics - Big Data - Data Lake					Align BOT to support future growth
Need to Change Organization Structure							
6	Responsible for tone at the top, digital transformation Strategy	Owners	CEO	Operations Owner		IT	CEO and upper leadership publicly support implementation
7	How to execute	Integrated		Separated			Integrate BOT in execution
8	Change management	What we do to make money		Processes		Build Skills	Identify BOT owners
9	Build DTS competency	Internal		Consultants		3rd party	Train your leaders
Cash Flow Impacts							
10	Setting the budget	Year One		Year Two		Year Three	Ensure budget to support BOT build
11	Source of Cash	Internal		New Business		External	Tie savings or revenue growth to the sector that is utilizing the BOT

Software Development Approach

#							
	How will Technologies support my business via DTS						
1	What Role will IT play?	Enabler		Supporter			IT will set up and manage three environments; Development, TEST, Production
2	What Should our goal be for using Technology?	Innovator		Early Adopter		Follower	Make sure the software meets functional needs
2.1	What is our technology fear factor?	Too much risk?		Reduce risk to business?		Lots of business risk?	Software approach is a very safe approach
	Using DTS to create organizational value						
3	How much technology will we use to support our business?	Accounting & Rapt - Process automation	Operations	Accounts receivable	Supply chain	Application integration/Security	The software approach is very specific to a technology and not necessarily to a business tower
4	Generate revenue or reduce expense	Reduce labor expense	Data analytics	Automation		End-to-end integration	Software development normally is focused on point solutions
5	How to support or see future business	Analytics - Big Data - Data Lake					Data analysis must be incorporated into the build.
	Need to Change Organization Structure						
6	Responsible for tone at the top, digital transformation Strategy	Owners	CEO	Operations Owner		IT	Software builds normally focus on one business tower at a time
7	How to execute	Integrated		Separated			Software builds tend not to be integrated
8	Change management	What we do to make money		Processes		Build Skills	Software builds require specialized skill sets
9	Build DTS competency	Internal		Consultants		3rd party	Software builds require specialized skill sets
	Cash Flow Impacts						
10	Setting the budget	Year One		Year Two		Year Three	Software builds normally have multi-year payback
11	Source of Cash	Internal		New Business		External	Due to time to get up and running, most sources of funding will be internal and backward looking

In this chapter, we covered how to best approach implementation while considering your digital transformation strategy. In all cases, regardless of technology deployed, it must not be deployed in a vacuum or in/into a silo.

Each organization that uses technology to solve an issue will not reap the full benefits of that technology and most likely will be dissatisfied with what the technology delivers. What the business often experiences is a solution that provides one solution, at one point, but creates other issues for surrounding processes or people.

For those of you who doubt this approach, I would just say take an hour watching "Property Brothers" or "Fixer Upper" on HGTV. Each show focuses on the strategic need for a home, with all the criteria for the home set by the buyers. A budget is set, a location is chosen, and a plan or electronic rendition is produced to set the expectations of the buyer and the builders. The project starts and when complete, the actual product matches the agreed upon requirements.

What you are doing with your digital transformation strategy is no different. Without a strategy and a plan, how can you ever really get what you want or need?

In the next chapter, we take a very tactical look at project management as we give examples of what does and does not result in a successful project.

CHAPTER 6

Project Management – Making Your Strategy a Reality

Project management is a critical part of any implementation or change effort, and often the last on the mind of the project stakeholders. Without project management, a project is most likely doomed to failure or at the very least, will not meet the expectations of those who commissioned the project.

To help the reader prepare for a project manager, I will break my methodologies into three simple methodologies:

- The LEGO® methodology
- The "model airplane" methodology
- The "new home build" methodology

As you may guess, each methodology will vary in details and structure.

The LEGO® Methodology

If you have children or young siblings, you might be familiar with the famous LEGO® toys that have been loved and enjoyed by many since the 1930s. This toy illustrates the first methodology.

The LEGO® method is simple and practical. It takes pieces that have similar shapes and sizes and allows a user to fit together all that is required for what you want to build.

The Framework

Desired outcome	Who sets requirements	Results
Implement a new system	Person who pays for project	Implementation rarely meets buyers/owners expectations

Steps	Actions	Outcomes
Choose a product	Do some basic research and purchase software license	Usually pay above market rate
Choose implementation approach	Decide to internally implement software	Only very basic configurations are implemented
Align user requirements to software	Software is made available to users	Users don't use software and it becomes shelf-ware
Software does not meet owners needs	Complain to vendor that software is "not of any value."	License expires and dollars are wasted.

Now, I am not saying that the LEGO® method can never work. I am just saying that it normally does not work. Going back to my example, if your goal is to just build four walls of various colors, use your imagination and call it a castle, then that is just fine. But if you want to build the "Star Wars Universe Millennium Falcon," pictured below, you better have a plan.

Unfortunately, at many mid-market companies, this is THE approach. It is often hard to convince the buyers or stakeholders to invest more into the project management side of things, but for some reason, with all the information about project management, this remains a go-to approach.

The "Model Airplane" Methodology

My children are now adults, so I don't know how popular or relevant this example is to younger readers. Perhaps it is more for those of you who recall building a model airplane in days gone by. When I was a kid, I would spend hours at the hobby store selecting just the right model. I would look at jets, prop-driven aircraft and alien space craft. I would compare and contrast number of parts, colors required and complexity of the design. I would review what tools and supplies I had and confirm that my skills could meet the challenges. Often, I would speak with a store clerk and ask who built the last model and what challenges they had.

The "Model Airplane" Methodology

Desired outcome	Who sets requirements	Results
Implement a new system	Small team of stakeholders and project sponsor	Mostly successful, with application meeting most the needs of the team, but not all because complexity results in only specific stakeholders' needs being fully implemented

Steps	Actions	Outcomes
Understand overall business need driving need for new software	Respond to team complaints or requests for new software	Conduct some research and perhaps attend a conference
Review software	Conduct extensive research on what is the most recommended software to meet your needs	Reach out to software vendors and ask questions, resulting a lots of email and telephone traffic
Speak with peer groups	Reach out to friends and colleagues to get their opinion	Select a software meeting the need of a trusted source
Purchase software license	Obtain license and ask IT to load software	IT often refuses to load or it is loaded by software owner and may not fully function as intended
Choose who will implement software	Identify internal resources or external services provider	Team goes from load to build

Steps	Actions	Outcomes
Configure software	Configure software based on most vocal member of the team or based on project owner's opinion	Software does a great job for the stakeholders with the most influence
Roll out software	Communicate to the team the software is ready for use	Some of the team loves the software, others refuse to use it and say that it does not meet their needs

I am not saying this method does not work or results in complete project failures. What I am saying over the years where this method was deployed, the overall organization was not satisfied, and the services vendor normally was blamed for "things not working" as promised.

The "New Home" Methodology

The "new home" methodology is an end-to-end approach. When a person sets out to build a new house, there is a great deal of thought and planning - from budget considerations to design choices to the overall project management of ordering supplies.

Just a simple example...First, the budget has to be defined by the buyers, so everyone involved can gauge the scope of the project. This is one of the first issues I run into when looking at a project implementation. For some reason, CFOs like to keep budgets a secret from implementers. That is a bad choice, because one never really knows what is expected by the client.

Next, contracts have to be signed, plans need to be drawn up and agreed upon, permits have to be pulled, land has to be prepared, and supplies have to be ordered. If any of these items are missed, there will be and often is a downstream negative impact.

I won't bore you with the rest of the steps, but I think you get the idea. Building a new house is just not an off-the-cuff effort, but a very methodical and calculated effort.

The "New Home" Methodology

Desired outcome	Who sets requirements	Results
Implement a new system	All relevant stakeholders and users are involved	Most successful methodology. Normally meets the needs of all stakeholder groups and is viewed as a success and point of pride for all involved

Steps	Actions	Outcomes
Understand overall business need driving need for new software, process improvement or enhanced employee training	Quickly assemble project leadership team	Project has support from business leadership, roles and responsibilities are assigned
Develop research plan	Project team collaborates on solution	Project team works with internal and external experts to identify several best approaches
Budget approval	Project team and leadership team set budget	Entire team, vendors and service providers understand the rules of engagement
Select solution	Select the best solution	To avoid silos it is best to include a project management team for the selection of the solution. You promote a collaborative approach and are much more likely to meet overall business process needs and address employee capabilities and requirements. This will ensure the solution and software are aligned.

Steps	Actions	Outcomes
Project Charter	Author project charter based on agreed upon deliverables	Project charters will drive project success. It defines how to manage conflict, assigns roles and responsibilities and makes the entire project transparent to all involved/affected
Project Plan	Create project plan setting tasks, deliverables, and milestones	Project plans, like the architect's drawings, make the project real and move it from theory to action
Deliver project	Execute plans based on project charter and plans	A very successful project.

You may be concerned that it will be too costly. Well, I disagree. What is the cost of failure? What is the cost of purchasing software that is never used? Worse, what is the cost of disrupting your business for a project that is doomed to fail from the beginning?

Each methodology does have merit, but if a strategic, focused approach is leveraged, then using the "new home" methodology should prove no more costly than the other choices.

We have included a number of links to Project Management resources in the appendix and we would be glad to share insights upon request.

CHAPTER 7

Conclusion

Many years ago, I would take a course or read a book and not feel that that time spent was work the information obtained. I hope I met my goal, that you feel that your time reading this book was well spent.

We have covered many topics and approaches. But now you should better understand digital transformation strategies, and how to take a practical approach to implementing various strategies to drive continuous growth and profitability.

After all, business - even a non-profit business - is about positive cash flow. If a business is not generating positive cash flows, they won't be around too long. Recall what I said earlier in the book: If you can't identify what the technology will do for the company, then don't implement it.

A technology must be part of a solution, the solution must be part of a strategy, and that strategy must be integrated with the business needs, goals and aspirations.

Please feel free to contact me at steve.palomino@palscott.com if you have any questions or want to chat about my travels.

ADDITIONAL PROJECT MANAGEMENT RESOURCES

Project Management Institute (PMI)
https://www.pmi.org/

PMI membership:
https://www.pmi.org/membership/join

Project Management Software
https://www.microsoft.com/en-ww/microsoft-365/project/project-management

Teamwork
https://www.teamwork.com

Microsoft Teams
https://www.microsoft.com/en-us/microsoft-365/microsoft-teams

REFERENCES

1. Triumph of the Nerds. Public Broadcasting Service website. https://www.pbs.org/nerds. Accessed August 12, 2020.
2. Economic Time Series. Economagic website. http://www.economagic.com/em-cgi/data.exe/beapi/fips92. Accessed August 12, 2020.
3. MIS Quarterly Executive 15(2):123-139 · June 2016 [Page 2].
4. MIS Quarterly Executive 15(2):123-139 · June 2016 [Page 35, Table 6].
5. Ibid.
6. Kane G, Palmer D, Phillips AN, Kiron D, Buckley N. Strategy, not technology, drives digital transformation: becoming a digitally mature enterprise. Summer 2015 Research Report. Page 9
7. Ibid. Page 5.
8. MIS Quarterly Executive 15(2):123-139 · June 2016 [Page 35, Table 6].

Made in the USA
Monee, IL
05 November 2026

67953746R00045